HOW IT WORKS

Computers

by Rebecca Pettiford

BLASTOFF! READERS 3

BELLWETHER MEDIA • MINNEAPOLIS, MN

Blastoff! Readers are carefully developed by literacy experts to build reading stamina and move students toward fluency by combining standards-based content with developmentally appropriate text.

Level 1 provides the most support through repetition of high-frequency words, light text, predictable sentence patterns, and strong visual support.

Level 2 offers early readers a bit more challenge through varied sentences, increased text load, and text-supportive special features.

Level 3 advances early-fluent readers toward fluency through increased text load, less reliance on photos, advancing concepts, longer sentences, and more complex special features.

★ **Blastoff! Universe**

This edition first published in 2022 by Bellwether Media, Inc.

No part of this publication may be reproduced in whole or in part without written permission of the publisher. For information regarding permission, write to Bellwether Media, Inc., Attention: Permissions Department, 6012 Blue Circle Drive, Minnetonka, MN 55343.

Library of Congress Cataloging-in-Publication Data

Names: Pettiford, Rebecca, author.
Title: Computers / by Rebecca Pettiford.
Description: Minneapolis, MN : Bellwether Media, Inc., 2022. | Series: Blastoff! Readers: How it works | Includes bibliographical references and index. | Audience: Ages 5-8 | Audience: Grades 2-3
Identifiers: LCCN 2021049238 (print) | LCCN 2021049239 (ebook) | ISBN 9781644876008 (library binding) | ISBN 9781648346750 (paperback) | ISBN 9781648346118 (ebook)
Subjects: LCSH: Computers--Juvenile literature.
Classification: LCC QA76.23 .P484 2022 (print) | LCC QA76.23 (ebook) | DDC 004--dc23/eng/20211018
LC record available at https://lccn.loc.gov/2021049238
LC ebook record available at https://lccn.loc.gov/2021049239

Text copyright © 2022 by Bellwether Media, Inc. BLASTOFF! READERS and associated logos are trademarks and/or registered trademarks of Bellwether Media, Inc.

Editor: Betsy Rathburn Series Design: Jeffrey Kollock Book Designer: Gabriel Hilger

Printed in the United States of America, North Mankato, MN.

Table of Contents

What Are Computers?

Computers are machines that handle information. They run on **electricity**.

People use computers at work and at school. People also use computers for fun. They play games and watch movies.

CONTINENTS AND OCEANS

Many common items use computers. Some cars have computers to help people find their way.

Smartphones and **tablets** are types of computers, too. People can use them on the go!

How Do Computers Work?

Computers need **hardware** to work. Hardware includes the parts of a computer that people can touch.

Screens are hardware. Keyboards, printers, and speakers are hardware, too.

Parts of a Computer

Input hardware lets people enter **data** into computers. People type with keyboards. They click with mice.

People can even use **microphones** to tell computers what to do!

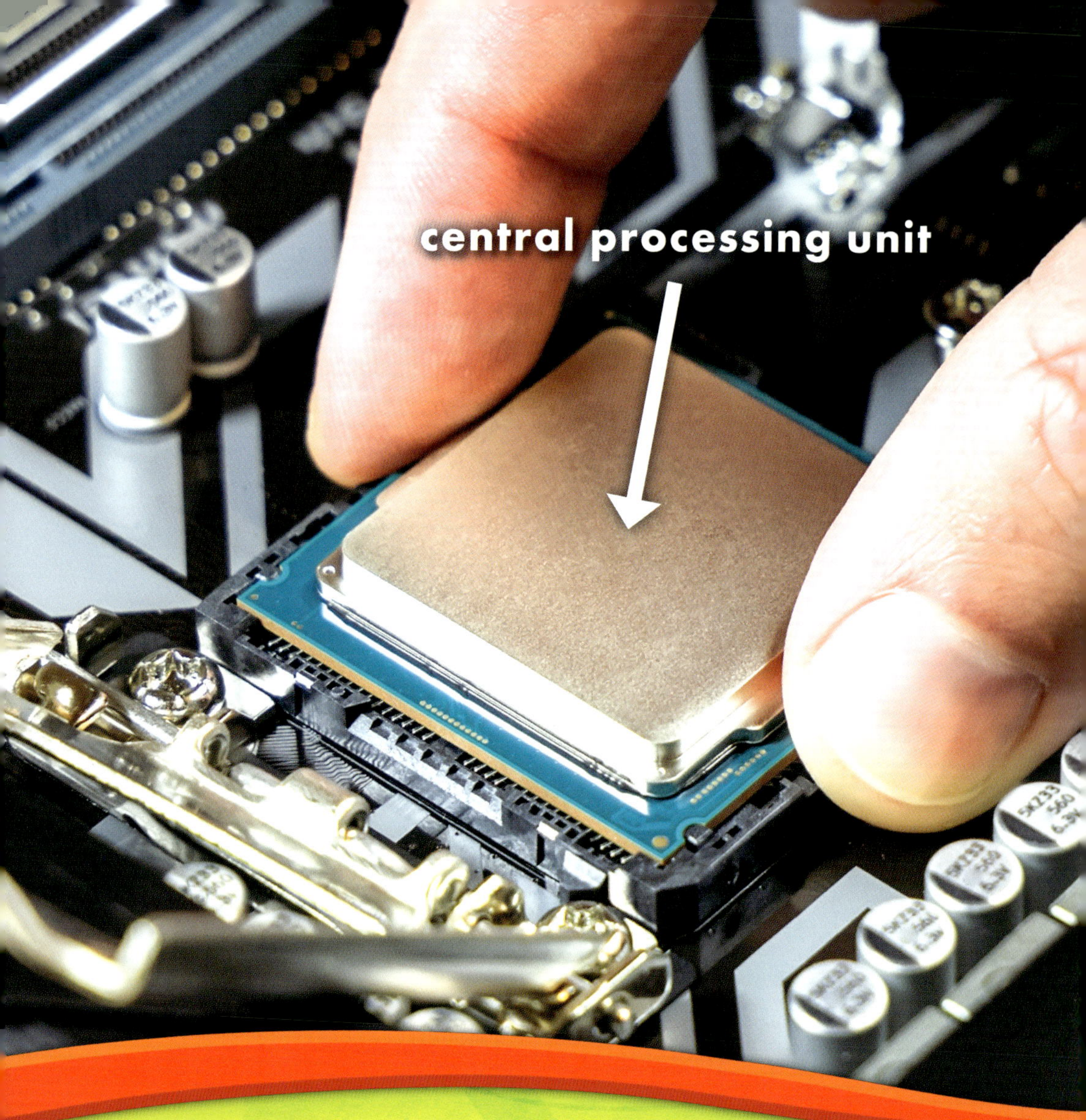

Data goes to the computer's **central processing unit**, or CPU. The CPU is hardware inside the computer.

How Computers Work

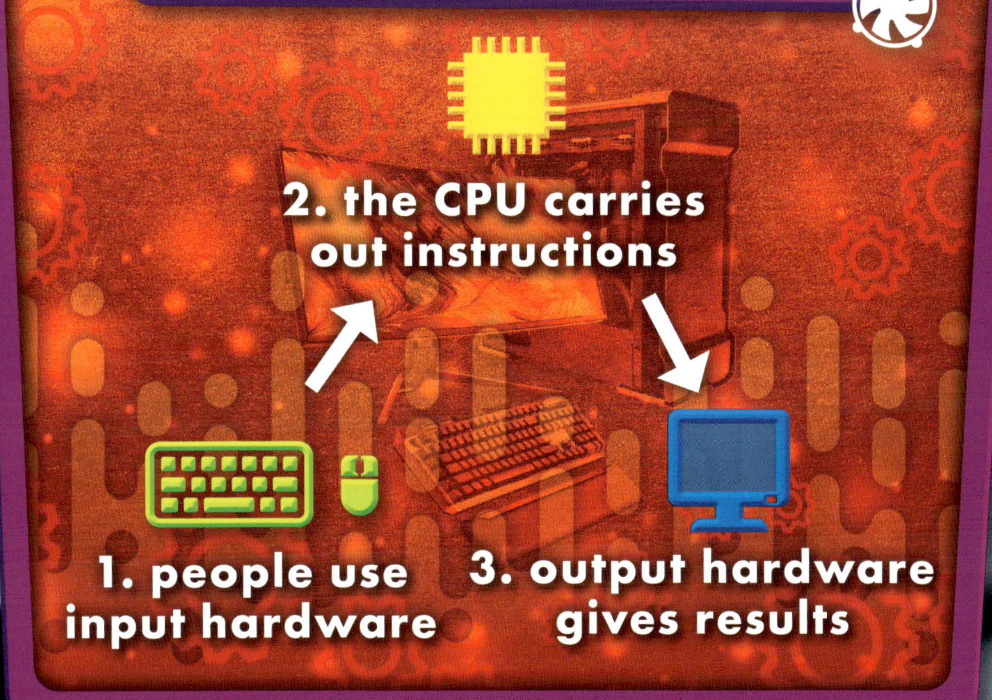

The CPU is like the computer's brain. It gets instructions. Then, it decides how to carry out the instructions.

The CPU sends the results to **output hardware**.

Speakers and printers are output hardware. But screens are the most important. They show the computer's **OS**.

EDUCATION
sharing
knowledge
school
design
creative
online
skill
successful
friends
connection
Experience
screen

The OS runs at all times on a computer. It makes using a computer easier!

Using a Computer

1

a person clicks to open an application

2

the CPU processes the data

3

the results show on the screen

People use the OS to open **applications**. People can listen to music, use the internet, and play games!

Computers need **memory**. It stores information. It helps them work faster.

RAM is a type of memory. It lets people do many things on their computers at once!

The Future of Computers

Future computers will look different. People could swallow tiny computers to find health problems.

Question

What do you think future computers will be able to do?

Computers will be more powerful. They will do their jobs faster. Future computers may change the world!

Glossary

applications–programs that do a certain thing, such as play music or visit websites

central processing unit–the part of a computer that does most of the processing of data

data–information

electricity–power that is carried through wires and used to run machines

hardware–the parts of a computer that people can touch

input hardware–computer hardware that lets people tell the computer what to do

memory–a device or system used to store data

microphones–input hardware that people can speak into

OS–operating system; the OS supports a computer's basic functions.

output hardware–parts people use to see the results of what they told the computer to do

RAM–random access memory; RAM stores information for a short time while an application is in use.

smartphones–cell phones with advanced features such as internet access and apps

tablets–portable computers

To Learn More

AT THE LIBRARY

Atkins, Marcie Flinchum. *Think Like a Computer*. New York, N.Y.: Children's Press, 2019.

Bethea, Nikole Brooks. *What is Coding?* Minneapolis, Minn.: Jump!, 2020.

Robinson, Peg. *How Computers Work*. New York, N.Y.: Cavendish Square, 2019.

ON THE WEB

FACTSURFER

Factsurfer.com gives you a safe, fun way to find more information.

1. Go to www.factsurfer.com.
2. Enter "computers" into the search box and click .
3. Select your book cover to see a list of related content.

Index

The images in this book are reproduced through the courtesy of: ApoGapo, front cover (keyboard); Kreabobek, front cover (computer); ArtoBarto, front cover (monitor); kvsan, p. 3; Gorodenkoff, pp. 4 (inset), 20-21; Rawpixel, pp. 4-5, 14-15; DenPhotos, pp. 6-7, 7 (inset); fizkes, p. 7; yafiq Adnan, pp. 8-9; LIAL, p. 9 (keyboard); Visit Roemvanitch, p. 9; Africa Studio, pp. 10-11; Odua Images, p. 11; hodim, pp. 12-13, 18-19; jannoon028, p. 14 (inset); PixieMe, p. 16 (step 1, step 3); William Potter, p. 16 (step 2); Lukmanazis, p. 17; JJFarq, p. 19; Lastroll, p. 23.